CARING FOR GOPHER SNAKE

GUIDE TO KEEPING GOPHER SNAKES AS PETS, CARE, FEEDING, AND MORE

DR MORRIS HART

Copyright© 2024 **DR MORRIS HART**

All rights reserved. No part or part of this book or publication may be reproduced, stored, or transferred in any form by electronic, mechanical, recording, or other retrieval system without written permission from the publisher

Table of Contents

INTRODUCTION 5

CHAPTER 1 11

How to Choose the Best Gopher Snake: A Complete Guide 11

CHAPTER 2 17

Creating the Ideal Environment for Your Gopher Snake 17

CHAPTER 3 25

Providing Food for Your Gopher Snake: A Complete Guide 25

CHAPTER 4 33

Managing and Getting Along with Your Gopher Snake: A Complete Guide 33

CHAPTER 5 43

A COMPLETE GUIDE ON GOPHER SNAKE HEALTH AND WELLNESS CARE 43

CHAPTER 6 55

COMMON QUERIES AND PROBLEM-SOLVING: AN ALL-INCLUSIVE HANDBOOK FOR GOPHER SNAKE KEEPERS 55

CHAPTER 7 67

CLOSING: SAVORING EVERY MOMENT WITH YOUR GOPHER SNAKE 67

Introduction

Bullsnakes, or gopher snakes, are interesting reptiles that have grown in popularity as pets in recent years. We will delve into the world of gopher snakes in this in-depth guide, examining their physical traits, temperament, and natural habitat—all of which contribute to why these reptiles make wonderful pets.

1. Knowing About Gopher Snakes

The genus Pituophis, which comprises many species indigenous to North and Central America, is home to gopher snakes. Because they are non-venomous constrictors, they squeak rather than splatter venom to immobilize their victim. Given their similar appearance, notably their patterned scales and triangular heads, gopher snakes are frequently confused for rattlesnakes.

Their more slender body and absence of rattles, however, set them apart.

2. Organic Environment

Gopher snakes are mostly found in a range of environments, such as woods, scrublands, grasslands, and deserts. They are incredibly adaptive animals that can live happily in a variety of settings, from lush forests to dry deserts. In their natural habitat, gopher snakes can be found lurking in vegetation, beneath rocks, or inside burrows. They seek small mammals, birds, reptiles, and eggs.

3. Physical attributes

Gopher snakes possess a variety of physical traits that allow them to flourish in their native environment. Their bodies are usually long and slender, with smooth scales that are available in a range of colors and patterns. Shades of brown, tan, yellow, and gray are common color variants, and they frequently have spots or stripes along their bodies. Gopher snakes also have characteristic facial characteristics, such as dark lines extending from their eyes to their mouths.

4. Conduct and Attitude

Gopher snakes are good pets for reptile aficionados of all skill levels because they are often quiet and gentle in nature. They are not likely to bite until provoked, though they may display defensive behaviors in response to threats, including as hissing, vibrating their tails, or striking postures. Gopher snakes can grow fairly docile and may even love being handled and investigated by their owners with consistent handling and interaction.

5. Why Invest in a Gopher Snake Pet?

Gopher snakes are great companions for reptile lovers for a number of reasons:

- Low Maintenance: Compared to other reptiles, gopher snakes require comparatively less maintenance, which makes them perfect for novices or people with hectic schedules.
- Long Lifespan: Gopher snakes can live up to 15 to 20 years in captivity with the right care, giving their owners years of companionship.
- Educational Value: For those who are curious about reptiles and their natural behaviors, owning a gopher snake as a pet can be a very beneficial educational experience.
- Interesting Behavior: Gopher snakes can be very entertaining to their owners because of their amazing diversity of actions, which include eating, hunting, skin-shedding, and exploring their surroundings.

6. Legal Aspects to Take into Account

It's crucial to learn about the local regulations around the ownership of reptiles before obtaining a gopher snake as a pet. Regulations pertaining to the ownership and maintenance of exotic pets, such as gopher snakes, may exist in certain areas. Before taking a gopher snake home, make sure you are aware of any permit needs, enclosure specs, or export/import limitations.

7. In summary

To sum up, gopher snakes are fascinating reptiles that provide a special and fulfilling experience of having a pet. Reptile fans of all ages can benefit greatly from gopher snakes because to its docile disposition, low maintenance requirements, captivating behavior, and stunning look. You can spend many years with your gopher snake if you give them a good home, eat well, and interact with them frequently.

Chapter 1

How to Choose the Best Gopher Snake: A Complete Guide

Making the proper gopher snake choice is essential to a happy and successful experience with reptile care. We will go over the many things to take into account when selecting a gopher snake as a pet in this extensive guide, including species selection, age and size considerations, health evaluation, and sourcing possibilities.

1. Selection of Species

It is imperative to educate oneself on the various species and subspecies of gopher snakes that are available for purchase before obtaining one. Other species, such as the Sonoran gopher snake (Pituophis catenifer affinis) and the Great Plains gopher snake (Pituophis catenifer sayi), may also be available, while the Western gopher snake (Pituophis catenifer) is the most popular variety kept as pets.

It's important to choose a species that fits your tastes and expertise level because different species may have slightly different temperaments and care needs. When choosing, take into account elements such adult size, coloring, and natural environment.

2. Age and Size Factors to Take Into Account

You must take the gopher snake's age and size into account while choosing one. Since they are often smaller and more vulnerable than adult gopher snakes, juveniles need extra care and attention to ensure healthy growth and development. Juveniles may be more enticing than adults because of their adorable appearance, but they may also need more frequent feedings and supervision.

However, because they are bigger and stronger as adults, gopher snakes are less vulnerable to stress and alterations in their surroundings. If you're a newbie or searching for a low-maintenance pet, they might be a better option. But, it's crucial to take the snake's size into account and make sure you have enough room for its enclosure.

3. Medical Evaluation

It's important to evaluate a gopher snake's general health and condition before buying one to make sure you're getting a robust and healthy pet for your home. Keep an eye out for symptoms of disease or injury, like:

- anomalies in the texture or color of the skin
- breathing problems or wheezing
- discharge coming from the mouth, nose, or eyes
- Sluggishness or inactivity

- unusual gait or motion

Furthermore, look for any lumps, bumps, or wounds on the snake's body that can point to underlying medical problems. Asking about the snake's eating history and any recent veterinary care it may have gotten is also a good idea.

4. Options for Sourcing

Pet stores, breeders, reptile expos, and internet merchants are some of the places you can find a gopher snake. Every choice has advantages and disadvantages, so it's critical to weigh aspects like reputation, dependability, and the condition and caliber of the snakes on offer.

Pet Shops: A variety of reptiles, including gopher snakes, are available at pet stores. Even if it's handy, make sure to examine the snake carefully and find out about its past and present care.

Breeders: If you're looking for healthy, socialized gopher snakes, reputable breeders are a great place to start. They can offer insightful advice on care and husbandry because they frequently possess a multitude of information and expertise.

Expos for Reptiles: These are gatherings of breeders and dealers where a range of reptiles and things linked to them are displayed and sold. They can be an excellent way to get to know knowledgeable breeders and view a large variety of gopher snakes.

Internet merchants: A few online merchants focus on selling supplies and reptiles. Online purchases of gopher snakes are easy, but you should do your homework to make sure you're dealing with a reliable seller and that the snake will be shipped responsibly and safely.

5. In summary

One of the most important steps in guaranteeing a happy and satisfying experience with reptile ownership is choosing the appropriate gopher snake. You may make an informed choice that satisfies your needs and tastes by taking into account variables including species selection, age and size considerations, health assessment, and source alternatives. Whether you select an adult or juvenile snake from a pet shop, breeder, or reptile expo, giving it regular vet care, a good diet, and a suitable home can help guarantee a long and healthy life for your new friend.

Chapter 2

Creating the Ideal Environment for Your Gopher Snake

For your gopher snake to be healthy, happy, and well-adjusted, its surroundings must be perfect. The main elements of a gopher snake enclosure will be covered in detail in this extensive guide, along with information on substrate selection, enclosure size and type, temperature and humidity requirements, lighting and heating options, hiding places and enrichment, and maintenance advice to keep your snake's habitat cozy and clean.

1. Size and Type of Enclosure

The first stage in creating a gopher snake habitat is determining the size and kind of enclosure that is suitable. Active reptiles, gopher snakes need plenty of room to roam around and investigate. It is advised to use an enclosure that is at least 40–50 gallons in size for a single adult gopher snake. But larger enclosures—75–100 gallons, for example—give your snake even more space to move about and get exercise.

Plastic tubs and glass terrariums are common choices for enclosure types. Plastic tubs are more affordable and require less upkeep than glass terrariums, which give superior visibility and aesthetics. Ensure that the type you select has a tight-fitting lid to stop leaks and preserve the right amount of humidity.

2. Selection of Substances

Creating a cozy and realistic habitat for your gopher snake in its enclosure requires careful consideration of the substrate you choose. Paper towels, coconut husk, cypress mulch, and aspen shavings are a few of the substrate possibilities available.

Because aspen shavings are inexpensive, absorbent, and simple to spot clean, they're a common material choice for gopher snake enclosures. For a more realistic burrowing experience for your snake, consider using coconut husk or cypress mulch. Steer clear of substrates like pine or cedar shavings, which can irritate your snake's respiratory system and release toxic chemicals.

Another substrate alternative is paper towels, especially for young snakes or those with respiratory problems. Paper towels are perfect for keeping a hygienic enclosure since they are simple to clean and replace, even though they don't look as good as other substrates.

3. Conditions for Temperature and Humidity

Your gopher snake's health and wellbeing depend on maintaining the right temperature and humidity conditions. As ectothermic reptiles, gopher snakes depend on outside heat sources to maintain their body temperature. Establish a temperature gradient inside the enclosure, with a warmer side that is between 85 and 90 degrees Fahrenheit (29 and 32 degrees Celsius) and a colder side that is between 75 and 80 degrees Fahrenheit (24 and 27 degrees Celsius).

You can use a combination of ceramic heat emitters, under-tank heating pads, and heat lamps to create these temperature gradients. To maintain ideal conditions, check temperatures frequently with a trustworthy thermometer and modify heating sources as necessary.

Another crucial element to take into account is humidity, particularly with regard to shedding and respiratory health. The habitat should have a 40–60%

humidity level, which can be attained by frequent misting, offering a humidity hide, or utilizing a humidifier specifically designed for reptiles. Using a hygrometer, check the humidity levels and make necessary adjustments to avoid too much moisture or dryness.

4. Options for Heating and Lighting

Despite not needing UVB sunlight like certain other reptiles, gopher snakes' behavior and biological cycles can be regulated by creating a natural day-night cycle with suitable lighting. Every day, set a timer for a full-spectrum LED or fluorescent lightbulb to offer 10–12 hours of light and 10–12 hours of darkness.

To maintain ideal temperatures inside the enclosure, more heating may be required in addition to lights. Effective heating alternatives include under-tank heating pads, ceramic heat emitters, and heat lamps. Place any heating elements outside the enclosure to save your snake from getting burned and to make sure it can't come into direct touch with them.

5. Secret Places and Enhancement

Gopher snakes are reticent animals that need lots of hiding places in order to feel comfortable and protected. Use a range of materials, such as cardboard boxes, half logs, and reptile caves, to create at least two hiding places—one on the warm side and one on the cool side of the enclosure.

To promote exercise and exploration, add branches, rocks, and other organic accents to your gopher snake's habitat in addition to hiding places. Use artificial plants to give visual interest and cover, and create climbing opportunities with robust branches or vines.

6. Upkeep Advice

Maintaining a clean and cozy habitat for your gopher snake requires routine care. Every day, give the enclosure a quick spot clean to get rid of any waste or soiled substrate. Every four to six weeks, give the enclosure a thorough substrate change.

To stop bacterial growth, replace water dishes with new, clean water every day and disinfect them frequently. To achieve ideal conditions, periodically check the temperature and humidity levels and make any required modifications.

Lastly, keep a frequent eye out for any indications of disease or stress in your gopher snake, such as alterations in behavior, look, or hunger. A long and healthy life for your pet gopher snake can be ensured by giving it the right husbandry and medical attention.

7. In summary

The health, happiness, and general well-being of your gopher snake depend on you providing it with the ideal home. You can create a comfortable and stimulating environment for your pet gopher snake to thrive in by choosing the right enclosure size and type, picking the right substrate, keeping the temperature and humidity levels appropriate, providing enough lighting and heating, providing hiding spots and enrichment, and adhering to routine maintenance tips. Your gopher snake will have many years of enjoyment and company in its new home if it receives the right care and attention.

Chapter 3

Providing Food for Your Gopher Snake: A Complete Guide

A balanced food is crucial for the growth, health, and general wellbeing of your gopher snake. Everything you need to know about feeding your gopher snake will be covered in this extensive book, including food requirements, feeding schedules, suitable prey items, feeding techniques, common feeding problems, and advice on how to keep up a healthy feeding schedule.

1. Knowing the Dietary Requirements for Gopher Snakes

As carnivorous reptiles, gopher snakes mostly eat small mammals, birds, other reptiles, and eggs in the wild. They use their strong constriction as ambush predators to subdue and devour their prey whole. For optimal nutrition and digestion, gopher snakes kept in captivity should be fed a diet that closely resembles the wild prey items.

2. Feeding Schedule

The key to keeping your gopher snake healthy and hungry is to set up a regular feeding plan. More frequent feedings are usually necessary for juvenile gopher snakes than for adults in order to sustain their rapid growth and development. Feed young snakes every five to seven days with suitable-sized prey pieces that measure around the same as the snake's widest point.

With their slower metabolisms, adult gopher snakes can eat less frequently—usually once every ten to fourteen days. To avoid underfeeding or obesity, keep an eye on your snake's physical condition and modify the feeding plan as necessary.

3. Suitable Prey Items

It's critical to select prey species for your gopher snake that are both appropriately sized and nutritionally suitable for its age and size. The most popular prey item for gopher snakes kept in captivity is frozen and thawed rodents because they are easily stored, easily obtainable, and very unlikely to harm your snake.

To guarantee safe ingestion and digestion, select prey items that are approximately the same width as the broadest point of your snake. Start with pinkie mice for young snakes, and as the snake becomes bigger, progressively increase the size of the prey. Larger prey, like adult mice or small rats, can be fed to adult gopher snakes.

4. Feeding Techniques

You can feed your gopher snake using a variety of techniques, such as the following:

Hand Feeding: Certain gopher snakes might feel at ease consuming food right out of your hand. Offer the prey item to your snake with tongs or forceps, letting it strike and constrict before it eats it.

Bowl Feeding: You may let your snake hunt and catch prey on its own by putting food items in a shallow dish or bowl inside the enclosure. This technique works especially well for reluctant feeders or hard-to-strike snakes.

Cage Feeding: If your gopher snake is larger, you can put the prey item just inside the cage and let it find and eat its own food. To guarantee that your snake safely eats the prey item, keep a close eye on the feeding procedure.

Regardless of the feeding technique you use, make sure someone is watching over the feeding procedure to avoid mishaps or injury. After feeding, take out any uneaten prey from the enclosure.

5. Typical Feeding Problems

Even while gopher snakes usually eat a lot, sometimes they have trouble eating or even refuse to eat. Typical feeding problems include the following:

Regurgitation: Inadequate temperatures, missized prey, or stress from handling could be the cause of your gopher snake's tendency to regurgitate its food soon after it has been eaten. Before giving your snake another meal, let it rest for at least a week. To avoid future regurgitation, make sure all the parameters for good care are met.

Refusal to Eat: Stress, disease, or alterations in its surroundings could be the cause of your gopher snake's prolonged refusal to eat. Examine the cage for any possible stressors, such as improper temperatures, insufficient hiding places, or excessive handling, and take appropriate action. See a veterinarian specializing in reptiles for additional assessment and advice if feeding problems continue.

6. Advice on Sustaining a Balanced Meal Schedule

Sustaining a nutritious feeding schedule is critical to your gopher snake's general health. The following advice can be used to make sure that feeding goes well:

Provide a comfortable setting for feeding: Reduce noise and distractions to provide a calm, stress-free feeding environment. To make your snake feel more comfortable and safe during feeding time, turn off any bright lights or loud noises.

Utilize suitable prey items: Based on the age and size of your snake, select prey items that are appropriately proportioned and nutritionally balanced. When possible, steer clear of feeding live prey because it puts your snake at danger for harm and isn't necessary for good nourishment.

Feeding activity should be closely observed by you to make sure your gopher snake is correctly striking, constricting, and consuming its prey. Take note of any anomalies or problems you find during feeding, and adjust as needed to avoid them in the future.

Maintain ideal husbandry conditions: Make sure the temperature, humidity, and hiding places in your snake's enclosure are all set properly to encourage a healthy digestive system and appetite. To maintain ideal conditions, routinely check on husbandry conditions and make any required modifications.

7. In summary

A balanced food is crucial for the growth, health, and general wellbeing of your gopher snake. You can make sure your gopher snake has a long and happy life in captivity by learning about its dietary needs, setting up a regular feeding schedule, providing suitable prey, employing safe feeding techniques, resolving frequent feeding problems, and keeping up a healthy feeding habit. Your snake will flourish on a healthy food and give you years of company and pleasure if given the right care and attention.

Chapter 4

Managing and Getting Along with Your Gopher Snake: A Complete Guide

Taking good care of your gopher snake and engaging with it is essential to responsible reptile keeping. We will go over all you need to know about caring for your gopher snake in this extensive guide, including handling methods, safety measures, socialization advice, and developing a close relationship with your pet.

1. Knowing the Behavior of Gopher Snakes

It's important to know your gopher snake's temperament and behavior before attempting to handle it. Gopher snakes are normally calm and inquisitive, but if they sense danger or stress, they may become defensive. Typical defensive actions include striking a striking position, shaking their tails, and hissing. You can steer clear of such confrontations and make sure that handling is enjoyable for both you and your snake by being aware of these indicators.

2. Safety Measures

When handling your gopher snake, safety should always come first. Take the following safety measures to reduce the possibility of harm coming to you or your snake:

To stop the spread of bacteria and illness, wash your hands both before and after handling your snake.

Avoid abrupt movements or rough handling of your snake since these actions may stress it out and even injure it. Instead, handle it carefully.

A good support system for your snake's body is essential to prevent injury and stress. Never grip or confine your snake by the head or tail.

Handling your snake while it is eating or digesting could cause regurgitation or other digestive problems.

To avoid mishaps or injury, keep an eye on how your snake interacts with other pets or family members.

3. Managing Methods

It's critical to handle your gopher snake with care in order to protect your safety and the comfort of your pet. To handle your snake safely and effectively, take the following actions:

Be confident and cool when approaching your snake; avoid making loud noises or abrupt movements that could frighten it.

Using one hand to support the front third of the snake's body and the other to support the middle and rear thirds, support its body with both hands.

Let your snake pass easily through your hands; do not squeeze or hold it tightly enough to make it uncomfortable.

Long-term handling of your snake can lead to tension and exhaustion, so try to avoid doing so. Ten to fifteen minutes at a time is the maximum amount of time you should spend holding your snake; as it gets more accustomed to being handled, you can extend the time.

Give your snake some time to calm down before trying to handle it again if it displays defensive behaviors like hissing or vibrating its tail. Whenever your snake seems agitated or uncomfortable, respect its boundaries and don't try to force an interaction.

4. Bonding and Socialization

Establishing a close relationship and developing trust with your gopher snake requires socialization. Here are some pointers to help you and your gopher snake become more social and bonded:

To assist your snake get used to handling and human interaction, handle it gently but frequently. As your snake gets more comfortable, progressively extend the length of your handling sessions from brief beginnings.

During handling sessions, it can help to reassure and comfort your snake if you speak to it in a soothing and quiet tone.

Give your snake lots of chances to explore and be enriched outside of its cage, such interactive playtime with toys and challenges or supervised free-roaming periods in a snake-proofed location.

When interacting with your snake, exercise patience and consistency, letting it grow at its own speed and being mindful of its comfort zones.

In order to positively reinforce excellent behavior and establish positive connections with handling and interaction, give goodies or rewards during handling sessions.

5. Establishing Trust

Although it takes time and effort to develop trust, it is crucial for forging a strong bond and having a good relationship with your gopher snake. The following advice can help you develop trust with your snake:

Treat your snake with care and gentleness; avoid abrupt movements or acts that could frighten or agitate it.

Give your snake lots of chances to explore and be enriched in a secure setting so it may become accustomed to its surroundings and gain confidence.

When your snake seems agitated or uncomfortable, don't force interactions or handling; instead, respect its boundaries and comfort zones.

Maintain consistency in your handling and socialization of your snake to help it get accustomed to human contact and develop trust over time.

During handling sessions, give sweets or awards to encourage good behavior and establish positive connections between handling and interaction.

6. Typical Problem Solving

Although caring for your gopher snake can be enjoyable, it's important to understand typical problems and how to solve them. Typical handling problems include the following:

Defensive behavior: Give your snake some time to calm down before trying to handle it again if it shows defensive actions like hissing, striking, or vibrating its tail. Be confident and calm when approaching your snake, and stay away from abrupt movements or acts that could frighten or agitate it.

Aggressive behavior: Gopher snakes rarely exhibit aggressive behavior, however it might happen in reaction to stress, disease, or other circumstances. For more assessment and advice, speak with a reptile veterinarian if your snake exhibits aggressive behavior.

Gopher snakes are expert evaders, and if given the chance, they can try to break out from their confines. Make sure the enclosure housing your snake has tight-fitting lids and sturdy closures to keep it from escaping.

7. In summary

Taking good care of your gopher snake and engaging with it is essential to responsible reptile keeping. You can make sure that handling your snake is enjoyable and fulfilling for both of you by learning about its behavior, taking safety precautions, handling it properly, bonding with it, and efficiently handling typical handling problems. You can create a strong friendship and have many years of fun and companionship with your gopher snake if you are patient, consistent, and respectful of its unique personality and comfort level.

Chapter 5

A Complete Guide on Gopher Snake Health and Wellness Care

A long and happy life in captivity for your gopher snake depends on maintaining its health and wellbeing. Everything you need to know about taking care of your snake's health and well-being will be covered in this extensive book, including routine health examinations, symptoms of sickness, preventive care options, common health problems, and advice on boosting general wellness.

1. Frequent Health Examinations

Monitoring the general health and well-being of your gopher snake involves doing routine health checks.

During health checkups, pay particular attention to the following areas:

Skin and Scales: Look for any anomalies, such as wounds, abrasions, or discolouration, on the skin and scales of your snake. The shiny, smooth scales of a healthy gopher snake should be devoid of blemishes and abnormalities.

Eyes: Examine the brightness and clarity of your snake's eyes. Sunken or dull eyes could be a sign of illness or dehydration, while cloudy or opaque eyes could be an indication of shedding or other health problems.

Mouth and Teeth: Gently open the mouth of your snake to look at the teeth, gums, and oral cavity. While healthy teeth should be spotless and devoid of any indications of decay or damage, healthy gums should be pink and wet.

Respiratory System: When breathing, pay attention for any unusual respiratory noises, such as clicking or wheezing. Breathing difficulties or a build-up of mucus around the mouth or nose may be signs of respiratory problems.

Body Condition: Look at your snake's general appearance and behavior to determine how healthy its body is. A gopher snake in excellent health should have a strong physique with well-defined muscles and plenty of energy.

Check your snake's health and behavior frequently, ideally once a week, to keep an eye out for any changes or irregularities.

2. Indices of Illness

For your gopher snake to receive a timely diagnosis and course of treatment, you must be able to spot the symptoms of illness. Typical symptoms of disease in snakes include:

Loss of Appetite: Stress, gastrointestinal disorders, respiratory infections, and other conditions can all be indicated by a loss in appetite or a refusal to eat.

Weight Loss: Excessive weight loss or discernible changes in physical appearance may be signs of underlying medical conditions, including organ malfunction, metabolic problems, or parasites.

Lethargy: In gopher snakes, unusual lethargy or inactivity may be an indication of disease or stress. A healthy snake should be aware of its environment, active, and attentive.

Respiratory Problems: In snakes, wheezing, coughing, or trouble breathing can be signs of respiratory infections or other problems. Keep a watchful eye on your snake's respiration, and if you observe any irregularities, get veterinarian attention.

Skin Problems: Deviations from normal skin tone or texture, such as redness, edema, or sores, could be signs of parasites, infections, or other dermatological conditions.

In order to receive additional assessment and treatment for your gopher snake, speak with a reptile veterinarian if you observe any symptoms of illness or anomalies.

3. Preventive Healthcare Practices

Maintaining the health of your gopher snake and delaying the emergence of frequent health problems require preventive treatment. Consider the following preventative care strategies:

Appropriate Nutrition: To guarantee that your gopher snake gets the nutrition it needs to stay healthy, feed it a balanced diet made up of small enough prey items.

Sufficient Hydration: Make sure your snake has access to clean, fresh water at all times, and keep a constant eye on its fluid levels. To stop bacterial development and dehydration, make sure water dishes are cleaned and replenished every day.

Environmental Enrichment: Provide your snake with an assortment of hiding places, climbing frames, and enrichment activities in its enclosure to provide it with ample opportunity for exploration, exercise, and cerebral stimulation.

Maintaining appropriate husbandry conditions is important for promoting general health and well-being. These variables include temperature, humidity, and lighting. To guarantee ideal conditions, routinely check the husbandry conditions and make any required modifications.

Frequent Veterinary Check-Ups: Arrange for your gopher snake to receive routine veterinary examinations so that you can keep an eye on its health and treat any possible problems early. To make sure your snake stays healthy, your veterinarian can do fecal testing, routine physical examinations, and other diagnostic treatments.

4. Typical Health Concerns

Despite being tough and durable in general, gopher snakes can nonetheless have some health problems.

Gopher snakes may suffer from a number of common health conditions, including:

Snakes frequently get respiratory infections, which are frequently brought on by bacterial, viral, or fungal diseases. Breathing difficulties, nasal discharge, coughing, and wheezing are possible symptoms.

Parasites: A variety of internal and external parasites, including internal worms, ticks, and mites, can harm gopher snakes and result in health issues. Managing parasite infestations requires routine fecal testing and parasite preventive strategies.

Problems with Shedding: Gopher snakes may experience dysecdysis, or improper shedding, as a result of low humidity, dehydration, or poor diet. Maintaining ideal husbandry conditions and giving a humid hide will ensure proper shedding.

Mouth Rot: A bacterial illness of the mouth and gums that can afflict snakes s known as mouth rot, also known as infectious stomatitis. Gum swelling, increased salivation, and oral lesions are possible symptoms. For mouth rot to be managed and complications from developing, prompt veterinary care is important.

Metabolic Bone Disease: Deficits in calcium and vitamin D lead to metabolic bone disease (MBD), a prevalent health problem in reptiles, especially snakes. Muscle weakness, soft, malformed bones, and trouble moving are possible symptoms. For gopher snakes to avoid MBD, proper diet and supplementation are crucial.

For an accurate diagnosis and course of treatment, speak with a reptile veterinarian if you think your gopher snake is having any health problems.

5. Encouraging General Well-Being

Encouraging general well-being is crucial to maintaining the health, happiness, and viability of your gopher snake in captivity. Here are some pointers for encouraging your snake's general well-being:

Establish a Comfortable and Enriching Habitat: Give your snake a cozy and stimulating habitat with the right substrate, hiding places, climbing frames, and environmental enrichment.

Provide Appropriate Nutrition: To guarantee that your snake gets the nutrition it needs to stay healthy, feed it a balanced diet made up of small-to medium-sized prey items.

Handle with Care: To reduce stress and help your snake get used to human contact, handle it firmly and carefully. Don't handle your snake roughly or too tightly, and be considerate of its comfort zone.

Regularly Check Your Snake's Health: Keep an eye on your snake's general health and wellbeing by doing routine health checks. Its behavior, appetite, and physical appearance should all be closely monitored. If you spot any irregularities or symptoms of disease, get it checked out by a veterinarian.

Provide Veterinary Care: To make sure your snake is healthy and to treat any potential health issues early on, schedule routine veterinary examinations. Your veterinarian can offer advice on how to maintain the best possible health for your snake through appropriate diet, husbandry, and preventive care.

6. In summary

Maintaining your gopher snake's health and well-being is crucial to its long and happy existence in captivity. You can help your snake thrive and have many years of companionship and enjoyment by doing routine health checks, identifying illness symptoms, putting preventive care measures in place, addressing common health issues, and promoting overall wellness through appropriate nutrition, husbandry, and veterinary care. Your gopher snake will continue to be healthy, content, and thriving for many years to come with the right maintenance.

Chapter 6

Common Queries and Problem-Solving: An All-Inclusive Handbook for Gopher Snake Keepers

Owners of gopher snakes may face a variety of queries and difficulties regarding the upkeep and conduct of their animals. We'll cover some of the most frequently asked topics in this extensive guide, along with helpful troubleshooting advice to get you past any obstacles you might run across.

1. When is the right time to feed my gopher snake?

The feeding pattern of gopher snakes varies based on their size and age. Compared to adults, juvenile snakes usually need more frequent feedings to sustain their rapid growth and development. Feed young snakes every five to seven days, providing them w th suitable-sized prey. With their slower metabolisms, adult gopher snakes can eat less frequently—usually once every ten to fourteen days. To avoid underfeeding or obesity, keep an eye on your snake's physical condition and modify the feeding plan as necessary.

2. If my gopher snake is refusing to eat, what should I do?

Gopher snakes frequently turn down food, particularly during shedding season or under stress. First, make sure that all husbandry parameters, including temperature, humidity, and enclosure size, are suitable if your snake is refusing to feed. To get your snake to eat, provide a range of prey items in all sizes and varieties. See a veterinarian specializing in reptiles for additional assessment and advice if feeding problems continue.

3. What symptoms might indicate a sick gopher snake?

Gopher snake illnesses can manifest as lethargy, odd skin or scale appearance, altered behavior, loss of appetite, and weight loss. For an accurate diagnosis and course of treatment, speak with a reptile veterinarian if you observe any symptoms of disease or anomalies in your snake. In order to effectively manage health conditions, early detection and action are essential.

4. What is the best way to determine when my gopher snake will shed its skin?

Gopher snakes may show a number of symptoms prior to shedding, including as dull or foggy eyes, changes in skin tone or appearance, an increase in hiding, and a decrease in hunger. Make sure the enclosure has the right amount of humidity to encourage shedding, and provide your snake a warm spot to hide out while it sheds. To avoid stress or harm, don't handle your snake during shedding.

5. In the event that my gopher snake escapes its cage, what should I do?

In the event that your gopher snake escapes its enclosure, be composed and thoroughly examine the surrounding area, keeping an eye out for possible hiding places like cracks, under furniture, and behind appliances. To keep track of your snake's whereabouts, sprinkle talcum powder or flour over doorways and other possible escape routes. To entice your snake out of hiding, make comfortable hiding places with heating pads or heat lamps. Look for any possible escape routes on your snake's enclosure and close them to stop it from escaping in the future.

6. Ways to deal with a gopher snake that is defensive?

When gopher snakes sense danger or stress, they may react defensively by hissing, vibrating their tails, or adopting a striking stance. Give your snake time and space to calm down if it is acting defensively before trying to handle it. Be confident and cool when approaching your snake; stay away from loud noises or abrupt movements. Avoid cornering or confining your snake; this could make it more defensive. Instead, handle gently.

7. Why does my gopher snake hide for so long?

Gopher snakes naturally hide because it gives them a sense of protection and security. Make sure your snake's enclosure has enough hiding places, like tunnels, logs, or greenery, if it spends too much time hiding. Make sure the temperature and humidity are within the proper range for the species of your snake. If your snake continues to hide, keep an eye out for any ndications of stress or disease, and get advice from a reptile veterinarian if needed.

8. How should my gopher snake's enclosure be cleaned?

Maintaining a clean and hygienic habitat for your gopher snake requires routine cleaning and upkeep of the enclosure. Every day, take out any leftover food or uneaten prey from the enclosure. Spot clean any soiled substrate as needed. Every four to six weeks, thoroughly clean the enclosure, replace the substrate, take out all the furniture and decorations, and sanitize it with a disinfectant appropriate for reptiles. Before putting fresh substrate in the enclosure and bringing your snake back, give it a thorough rinse and let it dry completely.

9. Is it possible to house several gopher snakes in one enclosure?

Although gopher snakes are typically solitary animals, some may be able to live together in a cage with appropriate individuals. Nevertheless, keeping several snakes in one enclosure raises the possibility of hostility, resource competition, and the transmission of infectious illnesses. If you decide to keep more than one gopher snake together, keep a close eye on their behavior for any indications of stress or violence, and be ready to separate them if needed. To reduce the likelihood of disputes, make sure the enclosure is big enough to comfortably house several snakes and has lots of hiding places and stimulation.

10. How do I handle an injured gopher snake?

In the event that your gopher snake gets a cut, abrasion, or bite wound, determine how serious the injury is and administer the proper first aid as required. To stop infection, clean the wound with a gentle antiseptic solution and treat it with a topical antibiotic ointment that is appropriate for reptiles. If you notice any swelling, redness, or discharge, it's important to constantly monitor the injury for signs of infection or complications. If required, get further examination and treatment from a veterinarian specializing in reptiles.

11. Why is the skin of my gopher snake faded or dull?

In gopher snakes, dull or discolored skin could be a sign of approaching shedding, dehydration, stress, or underlying medical conditions. Make sure the enclosure has the right humidity levels to encourage shedding, and if needed, give your snake access to a shallow water dish for soaking. Make that the lighting, humidity, and temperature in your snake's habitat are all within the proper range for its species. For additional assessment and treatment, speak with a reptile veterinarian if skin problems develop or continue.

12. How can I detect stress in my gopher snake?

Reduced hunger, hiding habits, defensive postures, excessive pacing or rubbing against cage walls, and unusual pigmentation or skin changes are some indicators that a gopher snake is under stress. Determine and take care of any environmental stressors that might exist, such as insufficient hiding places, unsuitable temperatures, or intensive handling. To reduce stress, give your snake a calm, safe habitat, and keep a close eye on its behavior to see if it starts to improve.

13. If my gopher snake regurgitates its food, what should I do?

Gopher snakes may regurgitate for a number of reasons, including as the wrong size of prey, insufficient temperature, or handling stress. Before giving your snake another meal if it regurgitates its previous one, take out any leftover food from the cage and let it rest for at least a week. Make sure the husbandry conditions are within the proper range for the species of snake you own, and keep a close eye out for any indications of illness or stress in your pet. See a veterinarian specializing in reptiles for additional assessment and advice if regurgitation continues or recurs.

14. What symptoms might indicate an underweight or overweight gopher snake?

You may determine if your gopher snake is overweight, underweight, or at an optimal weight by routinely checking its bodily condition. A gopher snake in excellent health should have a strong physique with well-defined muscles and plenty of energy. Excessive fat deposits, low muscular tone, and visible ribs or spine are indicators of being overweight or underweight. To keep your snake in a good physical condition, adjust the size of your prey and the frequency of feedings. If you are unsure of your snake's weight, seek advice from a reptile doctor.

15. Why is the mouth of my gopher snake hanging open?

In gopher snakes, mouth breathing or gaping may be an indication of respiratory problems, such as pneumonia or respiratory infections. Keep a watchful eye out for any further respiratory d stress indicators in your snake, such as coughing, wheezing, or nasal discharge. For an accurate diagnosis and course of action, speak with a reptile veterinarian. Maintain respiratory health by providing the right husbandry settings, such as the right temperature and humidity levels, and refrain from touching your snake excessively if it seems anxious or ill.

In summary

Owners of gopher snakes may face a variety of queries and difficulties regarding the upkeep and conduct of their animals. You can take care of any problems that may come up and give your gopher snake the best care possible by being aware of frequently asked questions and helpful troubleshooting advice. Always keep an eye on your snake's health and behavior, give it the proper husbandry circumstances, and get medical attention if you have any worries about its welfare. Your gopher snake will flourish and provide you with years of entertainment and company if given the right care and attention.

Chapter 7

Closing: Savoring Every Moment with Your Gopher Snake

Congratulations on obtaining a gopher snake and being the proud owner of one! You will have a fulfilling and enlightening experience as you set off on this rewarding and enriching path of reptile keeping. We'll discuss the benefits of living with a gopher snake in this conclusion, along with some parting remarks to help you get the most out of your time with it.

1. Creating a Solid Bond

The chance to have a close relationship and bond with your gopher snake is one of the most satisfying elements of owning one. You will build a trusting bond with your snake via consistent handling, care, and interaction that will only get stronger with time. Cherish the times you spend with your snake and take the time to observe and comprehend its own personality and mannerisms.

2. Acquiring Knowledge and Development

Taking care of a gopher snake requires ongoing learning and development. You will learn a great deal about the behavior, health, and husbandry of reptiles while you take care of your snake. Seize the chance to learn more about these amazing animals and develop a greater respect for the natural world.

3. Respecting the Beauty of Nature

With their brilliant patterns, sleek bodies, and distinct personalities, gopher snakes are aesthetically pleasing and enthralling animals. Give your snake some time to enjoy its inherent beauty and the wonder of the environment it lives in. There's always something breathtaking about these magnificent reptiles, whether you're watching them slither elegantly across their enclosure or cuddling up to them under their heat lamp.

4. Talking About the Experience

Having a gopher snake may also be a fun and fascinating shared experience with friends, family, and other reptile lovers. Show off your love of snakes to others by sharing images, tales, or opportunities for education. Promote greater knowledge and respect for reptiles in general by encouraging others to learn about and enjoy these amazing animals.

5. Accepting Accountability

The satisfaction of having a pet also entails having a gopher snake and taking care of it properly. Take full responsibility for this duty and make every effort to meet your snake's environmental, emotional, and physical needs. You can build a happy, fulfilling life for yourself and your pet snake by putting their wellbeing first.

6. In conclusion, a voyage of exploration

To sum up, having a gopher snake as a pet is an adventure filled with exploration, camaraderie, and awe at the wonders of nature. As you set out on this adventure, enjoy every second you spend with your snake and welcome the challenges and rewards that come with owning a reptile. Your gopher snake will enhance your life in ways you never would have thought possible, whether you are admiring their beauty, picking their brains about behavior, or just enjoying their company.

Thus, as you proceed on this journey with your gopher snake, keep in mind to treasure the times, welcome the new experiences, and relish the voyage. You two will make a lifetime of memories and establish a relationship that will last a lifetime. Cheers to many more years of joy and friendship with your cherished gopher snake.

www.ingramcontent.com/pod-product-compliance
Lightning Source LLC
Chambersburg PA
CBHW050239230526
45470CB00005B/2023